Computing and Systems Technology Division 2019

AA008736

Held at the 2019 AIChE Spring Meeting and 15th Global Congress on Process Safety

New Orleans, Louisiana, USA
31 March - 4 April 2019

ISBN: 978-1-5108-8695-7

Printed from e-media with permission by:

Curran Associates, Inc.
57 Morehouse Lane
Red Hook, NY 12571

Some format issues inherent in the e-media version may also appear in this print version.

Copyright© (2019) by AIChE
All rights reserved.

Printed by Curran Associates, Inc. (2019)

For permission requests, please contact AIChE
at the address below.

AIChE
120 Wall Street, FL 23
New York, NY 10005-4020

Phone: (800) 242-4363
Fax: (203) 775-5177

www.aiche.org

Additional copies of this publication are available from:

Curran Associates, Inc.
57 Morehouse Lane
Red Hook, NY 12571 USA
Phone: 845-758-0400
Fax: 845-758-2633
Email: curran@proceedings.com
Web: www.proceedings.com

TABLE OF CONTENTS

550371 Development of a Chemical Accident Response Scenario for Central Accidents Headquarters in Korea

The regional disaster countermeasures headquarters response accidents through 13 management units in the event of a chemical accidents and establishes an integrated support center to support and direct the on-site emergency response control team. However when the scale of the chemical accidents is too large for the regional disaster countermeasures headquarters, cooperation with the central government or other municipalities is needed. In this study, a disaster response scenario which is focused on chemical accident is developed for central accidents headquarters in connection with the regional disaster countermeasure headquarters. 5 management units including the situation analysis/information management, the related organization cooperation unit, the effect investigation team, the media response unit, and the complex incident response unit are selected and their tasks are defined.

The scenario consists of a timeline of 4 steps and the responses of each management units. And the scenario is based on the assumption that the accident is solved smoothly. In order to apply the scenario to real training, the 7 major problems of the central accidents headquarters are selected. These major problems is linked to the main issues of the existing regional disaster countermeasure headquarters scenarios and a framework to effectively response to each major problems is developed. This framework consists of 9 steps including recognizing the situation, assigning for a task to management units, determining how to response the situation, and so on. Evaluation method for countermeasure behavior is proposed to measure the suitability of trainees' responses to the situation. It is expected that the integrated scenario for the central accidents headquarters developed in this study will enable the integrated training of municipalities and the central government.

Copyright © American Institute of Chemical Engineers. All rights reserved.

Closed-Loop PID Re-Tuning in a Digital Twin by Re-Playing Past Setpoint and Load Disturbance Data

Junho Park [1], Cameron Patterson [1], Jeffrey D. Kelly [2], John D. Hedengren [1]

[1] Brigham Young University, UT, USA
[2] Industrial Algorithms Limited, ON, Canada

Keywords: PID tuning, exhaustive search, temperature control, digital twin, cyber-physical, smart tuning

Introduction

The PID controller is the most widely used basic regulatory control algorithm. PID control is important in chemical engineering processes as it plays a critical role to form the basis of advanced process control and optimization systems such as model predictive control (MPC) and real-time optimization (RTO). However, its performance can vary greatly on the tuning of its three (3) parameters. There are several different types of PID tuning rules or heuristics reviewed in [1]. Unfortunately, it can be demanding for operators, instrument technicians, process engineers, or inexperienced process control engineers to choose the most suitable rule and use the rule properly in the actual or physical operating system or environment. However, this demanding exercise is facilitated by excellent automated tuning selection within Distributed Control Systems (DCS), Programmable Logic Controllers (PLC), or through specialized tuning software. Tuning rules are valuable as a starting point for further manual tuning. Optimization-based PID tuning is the other option and, thanks to the advances in computational power, can be a viable approach provided the objective function and setpoint / load disturbance scenarios are properly managed. Numerical optimization techniques are divided into two different categories. One is a model gradient-based optimization and the other is non-gradient based or derivative-free optimization. However, because of the non-convexity nature of the PID tuning model, a majority of the PID tuning research has been done using non-gradient optimization algorithms such as particle swarm [2, 3], and extremum seeking [4] algorithms. Another group of research study has been done using random search methods or meta-heuristics including genetic algorithm [5], for example.

The main objective of this study is to propose a practical technique for using past or historical setpoint and load disturbance data to re-tune the PID controller's installed tuning parameters in a more effective way. The brute-force or exhaustive search method guarantees an optimal value of the objective function without the risk of stalling at a local minimum at the expense of traversing a pre-defined PID parameter search space or region. Additionally, the reliability of the tuning result is further increased by setting the simulation conditions to be as close as possible to the real or actual operation to create a digital twin or cyber-physical view of the installed characteristics of the PID controller.

Rule-Based PID Tuning

Rule-based PID tuning methods can be divided into two categories: (1) process model from open or closed loop data and (2) closed-loop frequency and gain response analysis. Most of the open-loop tuning rules and many of the closed-loop tuning rules derive the PID tuning parameters by referring to a process model that is first order or second order plus dead time model (FOPDT or SOPDT). In this case, the

performance of the PID tuning is highly dependent on the process identification result. However, breaking the control loop might not be welcomed in a real operation and the process identification from the closed loop data may not be as accurate as the open-loop identification.
There are methods in the second category that refer to the frequency response of the closed loop data instead of constructing a closed-form process model. Zeigler-Nichols closed-loop tuning requires sustained oscillation data to obtain an ultimate gain (K_u) and ultimate period (P_u) [6]. To avoid driving a process to the limitation of the stability region to obtain the sustained oscillation data, a relay method is introduced by Åström and Hägglund [7]. Shamsuzzoha and Skogestad proposed the SIMC tuning method using closed-loop overshoot response data [8].

Overview of Method

This study proposes two main ideas to improve PID re-tuning practice in real operating environments. First, it uses an exhaustive search method, alternatively called a brute- force method. This method has several benefits. It is very simple to implement. It does not employ any of numerical methods other than a simulation of the expected installed performance by re-playing back past setpoint and load disturbance data. It simply evaluates all the possible combinations of PID parameters resulting in the list of the objective function values and finds the smallest or minimum value out of the list considering both output-error and input-move deviations, for example. The logic can be written in any computer programming language with just three (3) DO or FOR loops and may be easily parallelizable. Despite the simplicity of the method, its performance can be greater than the other advanced optimization methods, especially for the PID tuning because it never converges to a local minimum, stalling at a local stationary point in the search region. The second idea is that it employs several elements to pose simulation or surrogate conditions that better replicates the real, actual or physical operating environment of the PID controller by: (a) re-playing back the past or historical setpoint and load disturbances [9]; (b) allowing multiple, simultaneous and probability-weighted process models to be included in the digital twin simulations (i.e., multiple scenarios or situations each with their own probabilities) for robustness; (c) including multiple and simultaneous PID controller configuration formulations (Equation A or B) or even ad hoc controller designs; (d) specifying any type of performance objective function criteria i.e., simultaneously minimize the output-error and input-move variances, overshoot, etc. (e) adding stability rules in the search to cut-off unstable sections of the closed-loop operating space and (f) utilizing an indirect and constrained controller design technique as it does not directly invert the process model in anyway found in direct controller design methods and can be applied easily to nonlinear processes.

To test the utility of the exhaustive search method, an Arduino temperature control kit is studied. This is a physical device that has a heater and temperature sensor which can be controlled and measured. First, the temperature control kit is used to create the closed-loop operating data with de-tuned PID parameters and selected exogenous setpoint change sequences. A deterministic parametric process model is then identified using an ARX (AutoRegressive with eXogenous input) structured model using the GEKKO dynamic optimization suite [10] estimating coefficients using a least-squares prediction-error objective function. Then, the exhaustive search method evaluates the range or domain of the different P, I, and D parameters using starting with a centroid of the existing PID settings. The best search objective function found provides the P, I, and D settings which are updated to the corresponding PID parameters through the following equations found below. The PID controller is then run again with the temperature control lab now using the re-tuned PID parameters and the data recorded.

PID Equations

There are many derivations of PID formula rooted in the original continuous equation detailed in [1]. For implementing PID controllers in modern digital control platforms such as DCS's (Distributed Control Systems) or PLC's (Programmable Logic Controllers), two popular discrete forms are widely used in industry. One is the positional form (Eqn 1) and the other is the velocity form (Eqn 2), which are exchangeable.

$$OP_t = OP_{bias} + K_c[e_t + \frac{\Delta t}{\tau_I}\sum_1^t e_t + \tau_D \frac{PV_{t-1}-PV_t}{\Delta t}] \quad (1)$$

$$OP_t = OP_{t-1} + K_c[(e_t - e_{t-1}) + \frac{\Delta t}{\tau_I} e_t + \frac{\tau_D}{\Delta t}(PV_t - 2PV_{t-1} + PV_{t-2})] \quad (2)$$

where, $e_t = SP_t - PV_t$ and is the output-error.

Whereas the positional form calculates the controller output position (OP), the velocity form calculates the change in controller output (ΔOP). Although the positional form is more straightforward to understand as its P, I, and D terms are directly translated from its original continuous form, the velocity form has several advantages from the convenience perspective such as no additional logic is required for anti-reset windup [12]. In addition to the choice of the positional and the velocity form, there is one more derivation regarding the derivative kick that appears on the derivative terms in PID equations (Eqn 1, Eqn 2). By replacing the SP – PV in the derivative term with the PV, it reduces the chances of abrupt increment of the controller output when setpoint changes are made. In this study, the velocity form with the derivative kick free form is used.

Closed-Loop Identification and Estimation

The success of various PID tuning algorithms strongly relies on the accuracy of the process model. Notably, the IMC based tuning rules directly calculate the P, I, and D tuning parameter using the equations as a function of the process model (K, τ and θ). Additionally, this tuning rule requires an approximation of the process model to be of the first or second order plus dead time (FOPDT or SOPDT) form whereas this proposed method is agnostic to both the process and controller model forms. Furthermore, the accuracy of the process model can always be worse in closed-loop identification than open-loop identification. Thus, these tuning rules might be overly sensitive to the result of process model identification. The proposed method minimizes the chances of losing process information by using any type of higher order process model directly without requiring any additional approximation to the first order process model. In addition, this new proposed method uses closed-loop routine operating data (with feedback) and may fit ARX process models (Eqn 3) although Box-Jenkins models may also be used. Both ARX and Box-Jenkins models have proven consistency in closed-loop identification [13, 14, 15]. The ARX form is

$$y_t = (a_1 y_{t-1} + a_2 y_{t-2} + \cdots + a_{na} y_{t-na}) + (b_1 u_{t-1-nk} + b_2 u_{t-2-nk} + \cdots + b_{nb} u_{t-nb-nk}) \quad (3)$$

where, na and nb represent the number of ARX coefficients that multiply by the same number of past output (y) and input (u) measurements. The parameter nk denotes the process time-delay or dead-time of the process excluding the inherent one time-period of sampling delay.

Replaying Setpoint and Load Disturbance

The proposed method uses the digital twin of the actual process to iteratively search for the optimal PID tuning. The ARX model identified from the closed-loop data is used as the base model of the digital twin. This identification may be distributed across multiple process models with appropriate chances of occurrence for a robust controller performance design i.e., including a family of process models. The next step is to evaluate the potential PID tunings through the digital twin by re-playing the same past setpoint and load disturbance (Eqn 4) as in the process data with

$$z_t = y_t - x_t \tag{4}$$

where, x_t represents the ARX model output for time-step t. This replay of the closed-loop data also describes the actual process more accurately.

Exhaustive Searching Method

To find the optimal PID parameters, the exhaustive searching method is used. The exhaustive search method simply scans all the possible PID tuning performances by evaluating the range of PID tuning combinations with the process digital twin. The upper and lower bounds of the P, I, and D tuning are set by the user. It can be set based on the current PID tuning as a re-tuning purpose or it can be tested for an unexplored range for an experiment. The computational time of the exhaustive searching method can be longer than the gradient-based optimization although it is easily parallelizable as each PID parameter set increment is independent from the other PID increments. However, the computational time is not a constraint of the PID controller tuning tasks as opposed to the real-time estimation problem that needs to return the solution faster than the pace of real-time. On the contrary, one of the major advantages of the exhaustive searching method is that it always guarantees a global minimum given the pre-defined search-space, unlike the other optimization methods. By considering the goal of PID tuning and the small size of this optimization problem, an exhaustive search has minimal computational time. If computational time were a concern, the exhaustive search could also be used to as a multi-start method in conjunction with a gradient-based optimizer to quickly find local solutions.

Two different types of objective functions may be considered for PID tuning (Eqns 5, 6). The objective functions are variations of the PID control performance indices known as ISE (Integral Squared Error) and IAE (Integral Absolute Error). The objective function consists of the output-error (OE) term which appears in ISE and IAE, and the input movement (IM) term. The optimization solution of ISE combined with input movement (or, rate of change) has been analytically derived and investigated in [11] and is the simplest form of move suppression. These multi-objective functions can be express in two different ways. One is Archimedean and the other is the lexicographic form (or goal programming), which shown in Eqns 5, 6 and Eqns 7, 8, respectively.

$$L_2 - norm: \quad \min_{Kc,\tau_I,\tau_D} J = \sum[w_{OE}\{SP(t) - x(t)\}^2 + w_{IM}\{OP(t) - OP(t-1)\}^2] \tag{5}$$

$$L_1 - norm: \quad \min_{Kc,\tau_I,\tau_D} J = \sum[w_{OE}|SP(t) - x(t)| + w_{IM}|OP(t) - OP(t-1)|] \tag{6}$$

where, w is the weighting factor for each term in the objective function denoted in the subscription (OE and IM).

$$L_2 - norm: \quad \min_{Kc,\tau_I,\tau_D} J = \sum\{SP(t) - x(t)\}^2 \tag{7}$$

$$\text{Subject to, } \sum\{OP(t) - OP(t-1)\}^2 < UB_{IM}$$

$$L_1 - norm: \quad \min_{Kc,\tau_I,\tau_D} J = \sum |SP(t) - x(t)| \quad (8)$$

$$\text{Subject to, } \sum |OP(t) - OP(t-1)| < UB_{IM}$$

where, UB is the upper bound of the input movement (IM) which may be initially set by the centroid PID performance

Either the Archimedean or lexicographic form of the objective function can be used for PID controller tuning. In terms of convenience, the lexicographic form is easier to use because it requires one user input parameter, UB_{IM}, as opposed to the Archimedean form that requires two weighting factors on both OE and IM terms.

Case Study

Arduino Temperature Control Lab

To validate the advantages of the proposed method, the Arduino Temperature Control Lab (TCLab) is used [16] (Figure. 1). This pocket-sized device enables testing of various process control and optimization algorithms including model identification and model predictive control. The Arduino TCLab is adopted by many universities as a process control laboratory class exercise [17]. It consists of two sets of heaters, transistors, and temperature sensor that work as process, actuators, and measurements, respectively. The control algorithms can be written in MATLAB and Python in the computer and data is communicated via a USB serial port.

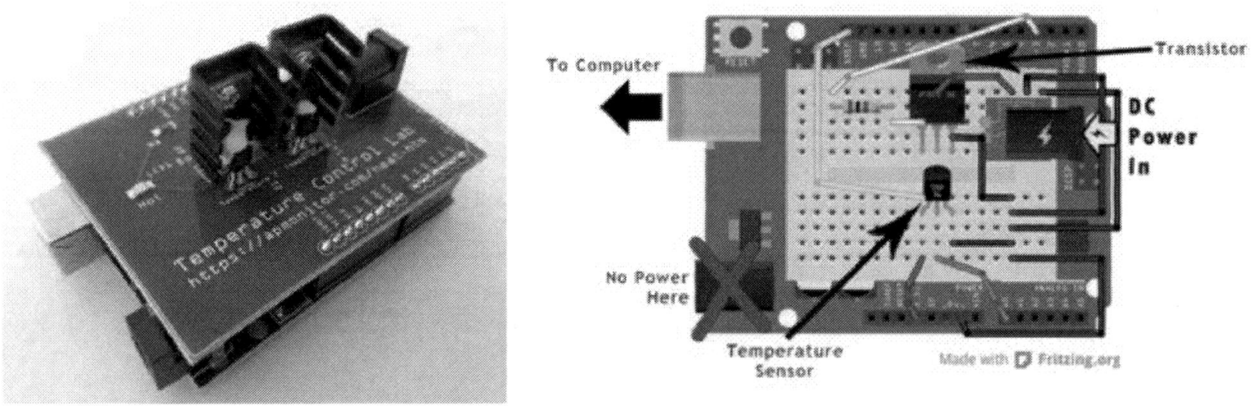

Figure 1. Arduino Temperature Control Lab (Photo and Schematic)

Results, Observations, Conclusions

The detailed results of every step of proposed method are shown in this section. The first step of the proposed method is to collect the closed-loop operation data and identify the ARX model parameters for identifying the digital twin. The setpoint is changed from 25°C at the initial steady-state condition to 35 °C. Figure 1 shows the measured temperature and ARX model fitting result on the same plot, and heater input value that is scaled between 0 to 100% on the right sub-plot. The ARX model parameters and the equivalent FOPDT parameter for the IMC tuning method are shown in Table 1.

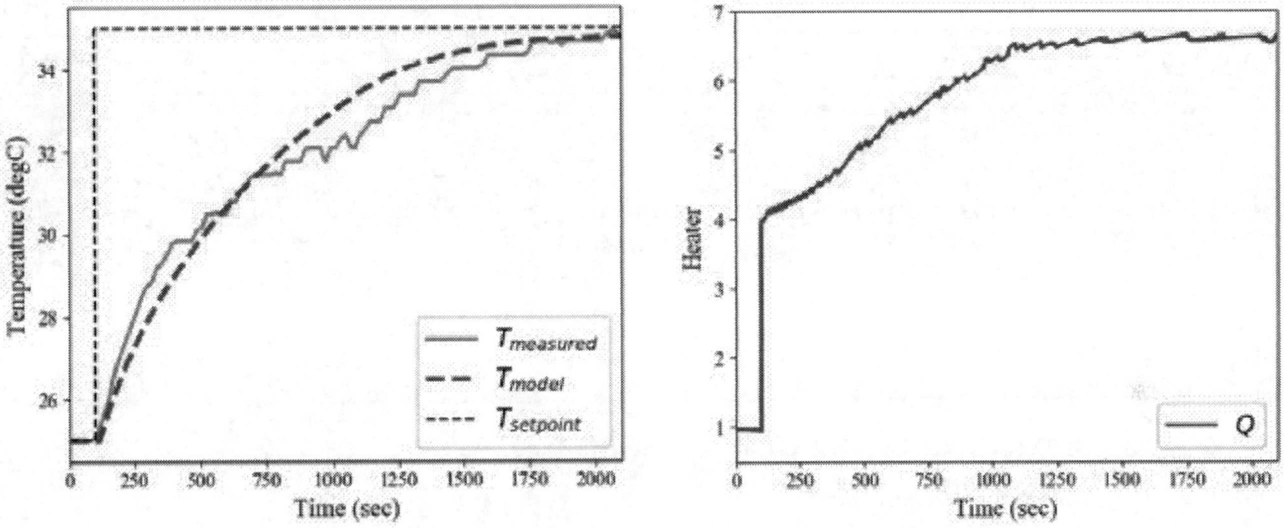

Figure 2. Closed-loop process data and digital twin model

For step 2, the range of K_c, τ_I, and τ_D is evaluated through the digital twin that includes the same setpoint sequences and load disturbance. The performance objective functions for each K_c, τ_I, and τ_D incremental combination are also calculated and stored. The L1-norm objective function in the lexicographic form is chosen for the experiment. The K_c, τ_I, and τ_D combination that gives a minimum value of objective function is then chosen as an optimal PID tuning suggestion. The simulation result of the suggested tuning is shown in Figure 3 along with the IMC tuning and the initial tuning results. The closed-loop time (τ_c) value is selected as one unit-of-time for this comparison. The IMC tuning result can also be faster or slower by setting a different τ_c, instead of setting the input movement (IM) limit for the proposed method.

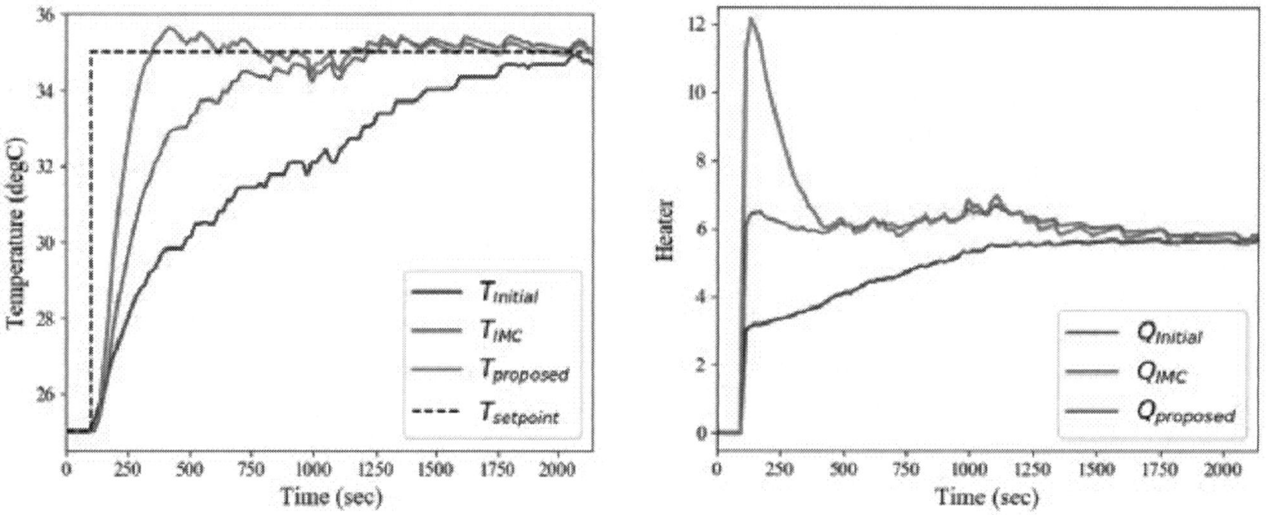

Figure 3. Improved tuning results suggested by exhaustive searching and IMC tuning (simulated with the digital twin)

Lastly, the suggested optimal tuning parameters are applied to the real system to validate the improved PID performance (Figure 4).

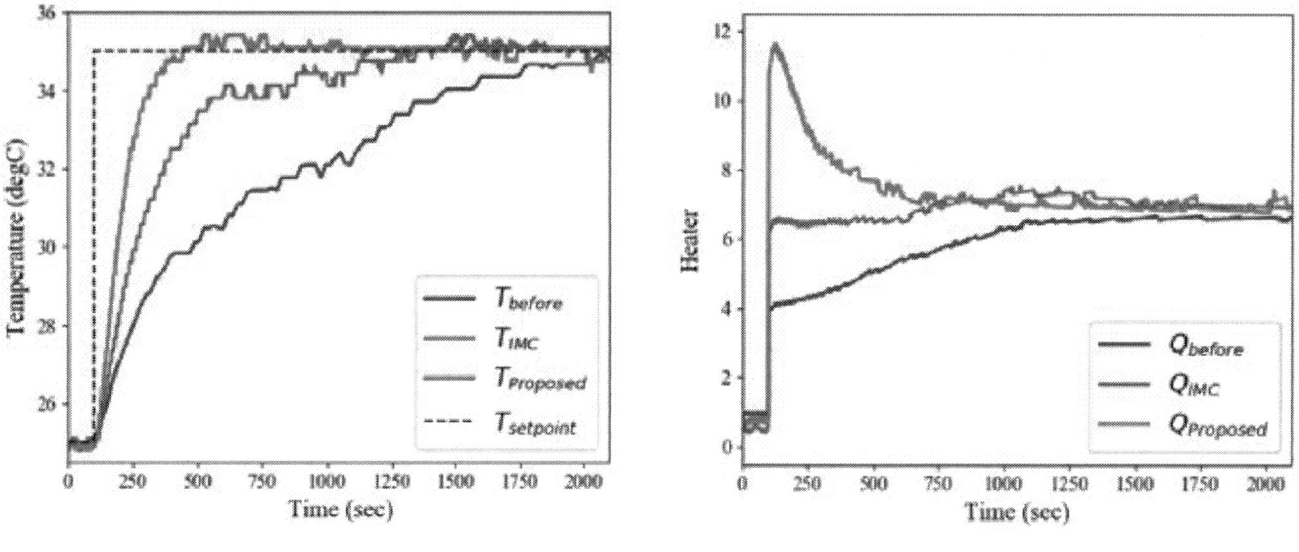

Figure 4. Performance validation in the physical system

Two additional case studies are shown in Figure 5 and 6 with different temperature ranges and different initial tuning parameters. The initial tuning for Case II (Figure 5) is set to generate the oscillating data, whereas Case III (Figure 5) shows sluggish initial control.

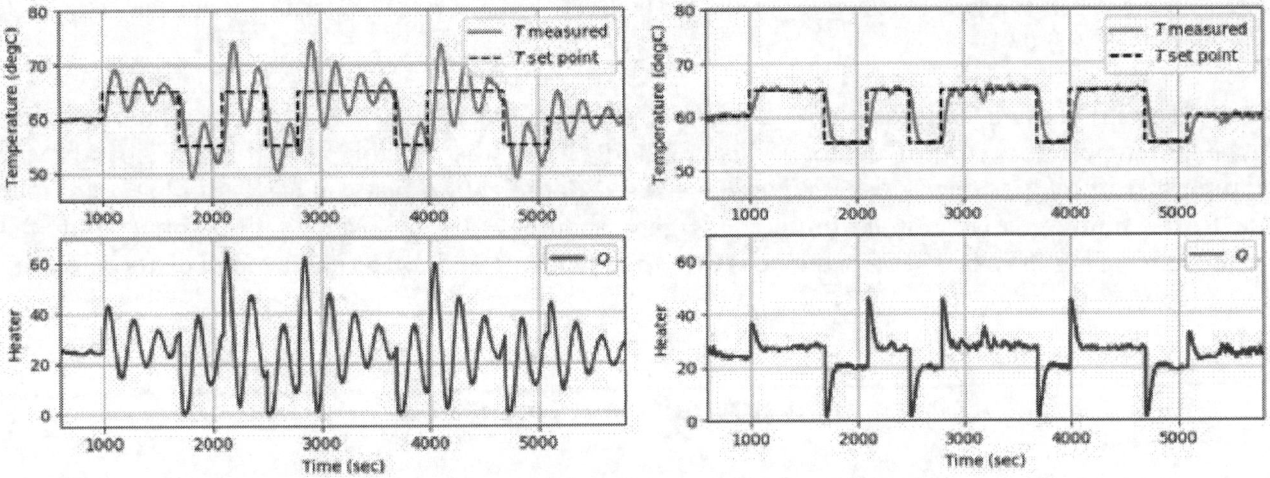

Figure 5. Case study II (Before and after tuning)

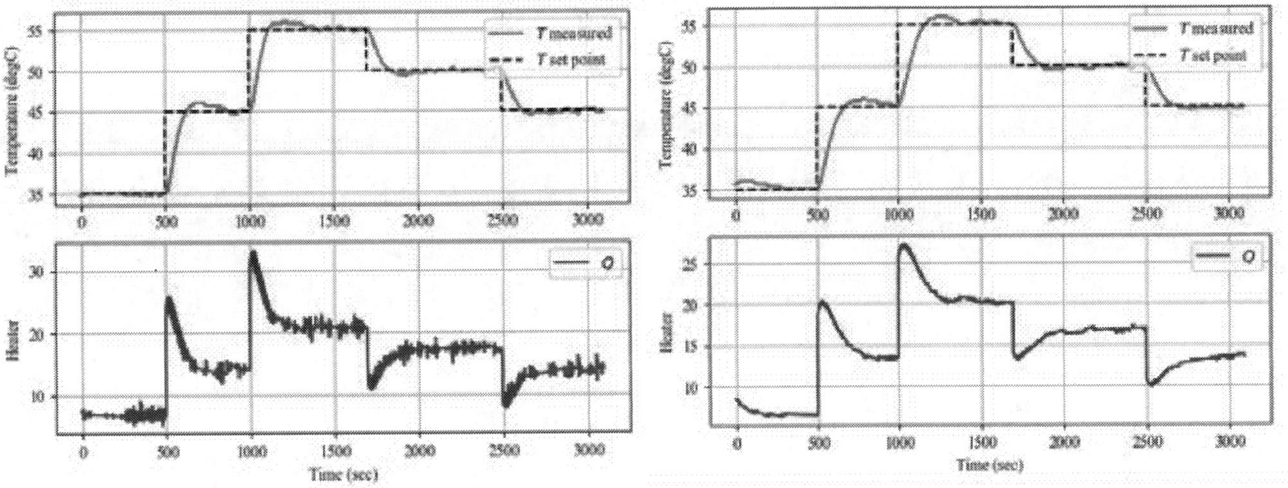

Figure 6. Case study III (Before and after tuning)

Table 1. ARX model parameters for case studies

	ARX Parameters		PID tuning						Performance Index			
			Before			After			Before		After	
	a_1	b_1	K_c	τ_I	τ_D	K_c	τ_I	τ_D	ISE	IAE	ISE	IAE
Case I	0.93	0.12	0.3	300.0	0.0	1.0 (0.56)*	150.0 (210.93)*	0.0 (0.0)*	27117	5690	6835	1434
Case II	0.90	0.20	1.5	18.0	0.5	2.0	450.0	0.1	71928	16072	42505	7833
Case III	0.92	0.12	1.7	100	5	1.2	100	0.64	13096	3716	13760	3287

Note: * refers the IMC tuning result ($\tau_c = 1$) and the FOPDT model parameters for IMC tuning are ($K_p/\tau_p/\theta_p =$ 1.78 / 210.93/ 0.0).

The PID tuning result using the exhaustive searching method is investigated and tested using the Arduino temperature control device. This method shows the reliable performance for the global minimum searching as opposed to the gradient-based optimization that can easily stall at local minima due to the non-convexity of the problem. Higher accuracy of the installed PID tuning potential is achieved by using the exact same sequences of setpoint change and reflecting the load disturbance during the process model evaluation.

Novel/Additive Information

- An exhaustive search method is applied to PID tuning suggesting a straightforward method that can be easily applied in real operation using a digital twin simulation and optimization.
- Re-tuning the PID controller in the digital twin and re-playing back past setpoint and load disturbances where the residuals of estimation are considered as the unmeasured load disturbances.

References

[1] Åström, K. J., & Hägglund, T. (1995). *PID controllers: theory, design, and tuning* (Vol. 2). Research Triangle Park, NC: Instrument society of America.

[2] Gaing, Z. L. (2004). A particle swarm optimization approach for optimum design of PID controller in AVR system. *IEEE transactions on energy conversion*, *19*(2), 384-391.

[3] Solihin, M. I., Tack, L. F., & Kean, M. L. (2011). Tuning of PID controller using particle swarm optimization (PSO). *International Journal on Advanced Science, Engineering and Information Technology*, *1*(4), 458-461.

[4] Killingsworth, N. J., & Krstic, M. (2006). PID tuning using extremum seeking: online, model-free performance optimization. *IEEE control systems*, *26*(1), 70-79.

[5] Mohanty, B., Panda, S., & Hota, P. K. (2014). Controller parameters tuning of differential evolution algorithm and its application to load frequency control of multi-source power system. *International journal of electrical power & energy systems*, *54*, 77-85.

[6] Ziegler, J. G., & Nichols, N. B. (1942). Optimum settings for automatic controllers. trans. ASME, 64(11).

[7] Åström, K. J., & Hägglund, T. (1984). Automatic tuning of simple regulators with specifications on phase and amplitude margins. Automatica, 20(5), 645-651.

[8] Shamsuzzoha, M., & Skogestad, S. (2010). The setpoint overshoot method: A simple and fast closed-loop approach for PID tuning. Journal of Process Control, 20(10), 1220-1234.

[9] Kelly, J. D. (1998). Tuning digital PI controllers for minimal variance in manipulated input moves applied to imbalanced systems with delay. *The Canadian Journal of Chemical Engineering*, *76*(5), 967-974.

[10] Beal, Logan, et al. "GEKKO Optimization Suite." Processes6.8 (2018): 106.

[11] Tchamna, R., & Lee, M. (2018). Analytical design of an industrial two-term controller for optimal regulatory control of open-loop unstable processes under operational constraints. ISA transactions, 72, 66-76.

[12] Seborg, D. E., Mellichamp, D. A., Edgar, T. F., & Doyle III, F. J. (2010). Process dynamics and control. John Wiley & Sons.

[13] Zhu, Y., Butoyi, F., & Dow Benelux, N. V. (2000). Multivariable and closed-loop identification for model predictive control. Technical paper.

[14] Voda, A., & Landau, I. D. (1994). Multi-step closed loop identification and control design procedure-applications, IFAC Proceedings, 27(8), 1543-1548.

[15] Jahanshahi, E., & Skogestad, S. (2013). Closed-loop model identification and PID/PI tuning for robust anti-slug control. IFAC Proceedings Volumes, 46(32), 233-240.

[16] Hands on Process Control, Hedengren, J.D., CACHE Newsletter, Computer Aides for Chemical Engineering, Dec, 2014, URL: https://bit.ly/2VfOJNH

[17] Rossiter, J.A., Jones, B.L., Pope, S., Hedengren, J.D. (2019). Evaluation and demonstration of take-home laboratory kit, 12th IFAC Symposium on Advances in Control Education (ACE 2019), July 7-9, 2019, Philadelphia, PA.

546659 Operational Experience with a Furnace Optimizing Tool for Burners Adjustment in Ethylene Cracking Furnaces and Refinery Fired Heaters in Order to Improve the Air-to-Fuel Ratio on Every Individual Burner, Increasing Both Efficiency and Safety and Lowering Emissions

Heide Refinery (HR) is a former Shell oil refinery located in the North of Germany. HR operates about thirty combustion units comprising fired heaters, boilers and incinerators. With increasing demands in respect to higher profit, higher efficiency and lower emissions (CO, NO_x, etc.), HR was obliged to improve operation of their fired equipment. One key problem was the adjustment of air flow to each individual burner so as to achieve uniform air-to-fuel ratios in each burner. This is a real challenge for all furnaces with a multiple burner configuration, because the air flow to each burner is normally not measured. Specifically during the start-up of furnaces, but also during normal operation, keeping the air-to-fuel ratio within a specific threshold range is crucial for safe operation. Furthermore, it allows achieving lower excess air levels without CO breakthroughs and, sometimes, also lowering NO_x emissions.

The adjustment of the individual burner air dampers and, therefore, of the air flows to each burner is normally the task of field operators. But operator experience differs from person to person and flame shape or appearance are judged differently in each case. Consequently, some burners operate closer to stoichiometric conditions than others. Operating these burners on lower excess air levels tends to generate higher flue gas temperatures and to increase the heat transfer to the product tubes. Over time, this causes overheating in some tubes or zones ("hot spots") and makes internal coke formation more likely. As a consequence, coke reduces the heat transfer, resulting in even higher tube surface temperatures. In ethylene cracking furnaces the tubes are normally monitored by manual pyrometer measurements on a daily or half-daily basis. The longer the running length of a furnace, the more likely it is that one or more tubes will exceed a critical surface temperature. When this happens, producers are forced to either reduce the feed flow rate or lower the feed outlet temperature. Both options normally lead to lower yield and lower total production. Finally, the furnace needs to be decoked. It is therefore an important goal of producers to heat the product tubes as uniformly as possible in order to achieve the longest running time between two decoking phases.

To improve the subjective individual burner air damper setting by the field operators, HR has equipped five furnaces (three ethylene cracking furnaces, the main crude oil distillation furnace and the vacuum distillation furnace) with a furnace optimizing tool that measures the air flow to each burner indirectly through the pressure drop that the air causes while flowing through the burner throats. Comparing the different pressure drops allows the operators to easily see which burner receives more or less air and to adjust the air damper accordingly. The tool is installed in furnaces with forced draft burners, although it also works perfectly well in natural draft burners.

Besides the need to reduce the hot spots, the three ethylene furnaces at HR had a severe air distribution problem. Air duct design to the 16 burners (per furnace) was poor and led to some burners operating on high excess air while others were starving on air, thus producing CO emissions. As emissions (NO_x, CO) are measured in the common stack, it was difficult to figure out which furnace and which burner were causing the CO problems. While CO was high, HR was not allowed to proceed increasing the feed throughput due to environmental permit restrictions. Therefore, CO had to be reduced first, and it took HR one to three days of intensive try-and-error seek until the problem was solved and throughput could be put to normal.

With the help of the furnace optimizing tool, this time was cut back to only 1-2 hours, leading to higher production and compliance with emission regulations at all times. In addition, the cycle time between two decoking phases increased, even though this was not the primary goal for these three specific furnaces.

The major goal of installing the furnace tool in the crude oil distillation unit (CDU) with 16 burners was not to prolong the running time between two decoking cycles, as CDU normally operates non-stop between 3 and 5 years. The challenge here was the fact that the CDU furnace operates on four different fuels at the same time: two different refinery fuel gases, a low-BTU off-gas and a heavy fuel oil. Maintenance requirements (e.g. of oil guns) had caused burners to operate on different heat releases. Burner air damper settings were based purely on visual flame observation, although the physical settings were kept almost always the same for all burners. It is obvious that different heat releases require different amounts of combustion air and, therefore, the air damper settings must be in accordance to the heat released by every burner. On top of that, air distribution throughout the air ducts was poor, leading to different air-to-fuel ratios in every burner.

By using the furnace optimizing tool, HR was able to adjust air-to-fuel ratio individually and to finally reduce the overall excess air of the furnace to levels as low as 1.5 Vol% O_2, with CO well below 10 ppmv and an improved flame shape.

On the vacuum distillation furnace, the goal was NO_x reduction. The furnace operates with four forced draft burners. Air distribution was also poor here. The optimizing tool allowed trimming primary and secondary air in every burner so that each became the same in every burner, even when they differed from each other. This has led to a NO_x reduction of ~22%, falling from 110 mg/Nm^3 (~54 ppmv) to 85 mg/Nm^3 (~41 ppmv) without any costly measures, just through burner damper adjustment.

The implementation of this furnace optimizing tool has multiple advantages over traditional operation. Depending on the specific needs and the equipment installed, it can achieve a shortening of cycle times between decoking phases (in ethylene cracking furnaces!), a reduction of the overall excess air with the subsequent increase in efficiency, a reduction of hot spots or flame impingement, a lowering of emissions, etc. Above all, operation becomes safer as burners can be kept within certain air-to-fuel ratio envelopes.

The investment is returned in 2 to 3 months. Furthermore, the tool can be installed and commissioned with the furnaces running, meaning there is no need for a shut-down. Operators are quickly convinced to use this tool when they realize that operation improves and they spend less time and make less adjustments in the furnace than before.

Copyright © American Institute of Chemical Engineers. All rights reserved.

551243 Multiphase Distribution and Holdup Investigations in Trickle Bed Reactors and Upflow Moving Bed Reactors Using Gamma-Ray Computed Tomography (CT) Technique

Multiphase flow reactors are gas-liquid-solid contacting equipment widely used in many fields such as petrochemical, petroleum, fine chemicals and biochemical industries. Trickle bed reactors, in which gas and liquid phase reactants flow in downward direction over a bed of solid catalyst particles, have quite versatile applications such as oxidation reactions, petroleum processing, hydrogenation reactions, esterification, F-T synthesis etc. Moving bed reactors are typically vertical reactors with gas and liquid flow upward through catalyst bed in hydrodesulfurization, hydrodenitrogenation, hydrodemetalization etc.

Phase distributions and holdups in multiphase flow reactors can affect many other design parameters such as wetting efficiency, heat and mass transfer coefficients etc. The phase holdup in the bed also controls the phase residence time and conversion of the reactants. It is therefore essential to understand how gas/liquid distribute at different axial levels and how gas/liquid holdup varies along such as (1) column diameter, particle size and shape, (2) operating conditions like gas/liquid flowrates, (3) physical and chemical properties of fluids.

Proper and highly accurate investigations of phase distributions and holdups are imperative for the design and diagnoses of multiphase flow reactors. Gamma-ray computed tomography is a non-invasive technique that provides the cross-sectional images at different axial levels by rotating the gamma source and its detectors around the object. It can visualize and quantify the phase distributions and holdup profiles for multiphase flow reactors which cannot be measured by other techniques. The Cs-137 source and its detectors are arranged and designed to provide a 40° gamma-ray fan beam with 5 mm height in the horizontal plane. This fan beam can cover columns up to 24 in. (0.6 m) in diameter for scanning. The detectors are shielded with lead collimators which has 5×2 mm fine apertures to obtain narrow gamma-ray beams and minimize the scattered gamma-ray in order to achieve better spatial resolution. The data acquisition system, which converts the detection of γ photons to the number of photon counts in computer, consists of NaI crystal detectors, preamplifiers, pulse processors and stepper motors. All the parameters such as the sample time, sample frequency etc. can be specified in the input text file which commands the motor controller to move the detector plates and source plate to a precise distance. While scanning, all the photon counts will be recorded in the output text file for future data analysis. For one CT scanning, Cs-137 source has 197 source positions (197 views). At each source position, the detector array moves counter clockwise for 20 times in 0.13° angular steps which are total 21 positions (21 projection measurements). Computer will collect the photon counts for each projection. After finishing the scan, the data of 62055 projections (197 views × 15 detectors × 21 projection measurement per detector) of gamma-ray that pass through the column will be written to the specified output text file. The main goals of CT data analysis are obtaining the phase distribution images (cross-sectional image for each level) and phase holdup profiles. The original data collected from CT scan are processed by alternating minimization (AM) algorithm to depict the attenuation values of the cross-sectional images.

Different multiphase flow reactors such as trickle bed reactor, upflow moving bed reactor have been deeply researched under many selected conditions using the Gamma-ray CT technique at Multiphase Flow Reactor Engineering and Application Laboratory (MFReal) in Missouri University of Science and Technology. The phase distributions at different axial levels can be easily obtained with high resolution images. Azimuthally averaged phase holdup profiles and local linear holdup profiles for different directions are the main achievements of Gamma-ray CT technique. Some other parameters such as wetting efficiency and internal liquid holdup inside porous catalyst are also able to be obtained combining with some special image processing methodologies.

Copyright © American Institute of Chemical Engineers. All rights reserved.

551272 Advancement Measurement Techniques for Validation of Computational Fluid Dynamics (CFD)

Multiphase flows and reactors are complex due to complex interactions of phases, thus design and scale-up of these reactors were least understood and mostly based on the empirical approach. Computational fluid dynamics (CFD) is theoretically sound and promising approach for the scale-up, design and process development of these reactors. However the development (in particular closures) and validation of CFD suffers from the lack of detailed experimental data which disseminates information about the local (spatial) mean as well as fluctuations data. Such a data needs sophisticated measurement techniques. In our laboratory, advanced measurement techniques like radioactive particle tracking (RPT), computed tomography (CT), gamma ray densitometry, optical fiber probes, heat transfer probes, four point optical fiber probe, two point optical fiber probe, liquid and gas tracer were developed/customized to obtain comprehensive and detailed data on different multiphase systems (gas-liquid, gas-liquid-solid and gas-solids) for advancing the knowledge and validation of CFD. Full flow field information of solid/liquid phases which includes Lagrangian flow field, local mean velocity, fluctuations velocity (in terms of RMS), turbulent quantities (stresses and kinetic energy) were obtained using Radioactive particle tracking (RPT). Computed tomography (CT) were used to obtain cross sectional phase distribution. Optical fiber probes were used to obtain bubble properties (bubble frequency, size, chordal length and interfacial area) and local heat transfer coefficient and mass transfer coefficient were measured from heat and mass transfer probes respectively. Global mixing and local mixing (diffusivity) were studied using tracer and RPT studies, respectively. Such a comprehensive data on these systems advances the hydrodynamics and heat transfer and their relationship. In this presentation, overview of these techniques and selected results using these techniques on the various multiphase systems like pebble bed reactor, fluidized bed, spouted bed, bubble column, trickle bed, upflow moving bed will be presented to disseminate the flow field, phase distribution, heat transfer, mass transfer and mixing, to demonstrate the detailed information available for the validation and development of CFD.

Copyright © American Institute of Chemical Engineers. All rights reserved.

544118 Incorrect Definitions and Uses of Tray/Packing Efficiencies in Process Simulator Codes

The Murphree efficiency concept (including Murphree component efficiencies and the Murphree tray efficiency, wherein all component efficiencies are equal) as well as the component vaporization efficiencies concept, were developed to improve upon the "equilibrium stage" assumption often used in the preliminary evaluation of a column's separation performance. For packed columns, the HETP (or NTSM) is used to as a link between equilibrium stage calculations and the column packed height. Even with the availability of powerful and fast personal computers and the development of rate-based column calculations which obviate the need for efficiencies altogether, Murphree tray/component efficiencies, component vaporization efficiencies, and HETPs are still used widely in the study of column performance. First, they often supply "good enough" estimates of column temperature and composition profiles. Second, they allow for faster and more stable column calculations compared to those based on rate-based theory and its numerical implementation. And finally, they are often the only option available for carrying out process calculations under some simulation conditions.

Unfortunately a number of well-known and widely used process simulators contain incorrect formulations of these quantities. In particular this talk will focus on the following issues:

Improper use of Murphree and vaporization efficiencies.
Ability to set "N" efficiencies when only "N-1" are independent.
Inclusion of constant tray or section vaporization efficiencies when such a concept is impossible.
Allowing $\Sigma \neq 1$
HETP for packed columns defined incorrectly for binaries and calculations for multicomponent systems are baseless.

Copyright © American Institute of Chemical Engineers. All rights reserved.

550602 Risk Analysis of Rare Events By Modified Hierarchical Bayesian Modeling (mHBM)

Risk assessment of rare but catastrophic events in process industry deals with challenges of data scarcity and uncertainty estimation. Data scarcity and uncertainty both are prevalent with operator responses as well as process data. Difficulty presented by sparse data for accounting uncertainty in failure probability of control layers can be solved by Bayesian model. Bayesian model networks when applied with fault tree, event tree *etc.* deals with the static nature of trees and lack of event occurrence data [1]. A fault tree connects basic events to initiating events and an event tree connects initiation event to consequences, rare events being one of them. However these trees themselves are unidirectional in nature [2]. Also, it is assumed that data collected for risk assessment comes from strictly consistent operating conditions. It leads to unaccountability of uncertainty associated with source-to-source variability in data sources for process data as well as operator responses. The research area of implementing bi-directionality in fault and event trees and estimating uncertainty due to source-to-source variability of data sources has not gained much attention for chemical process industry. According to literature [3], hierarchical Bayesian modeling (HBM) deals effectively with uncertainty presented by source-to-source variability of data sources in off-shore blowout.

The present work integrates the fault and event trees for chemical processes. Bi-directionality in the integrated model is achieved by means of Bayesian network. It enables top-down approach from consequences to basic events, hence making possible prediction of consequence probabilities by number of occurrences of basic events and vice-versa. The integrated model is amalgamated with HBM to deal with the factor that data points for process data and operator data comes from the conditions which may or may not be similar, tackling source-to-source variability and uncertainty in operator responses in process industry. The new technique is demonstrated on the case study of Tennessee Eastman problem (TEP). TEP is modified to include two more control layers of operator action and action effectiveness for consideration of human interaction with the process. Hence this work presents a novel integration model which includes a holistic method for development of a complete structure for risk assessment of rare events combining all sources of uncertainty.

Keywords: Hierarchical Bayesian modeling, fault tree, rare events

References:

1. He, Rui et al. "A quantitative risk analysis model considering uncertain information" Process Safety and Environmental Protection 118 DOI: 10.1016/j.psep.2018.06.029
2. Khakzad N, Khan F, Amyotte P. Dynamic safety analysis of process systems by mapping bow-tie into Bayesian network. Process Safety and Environmental Protection, 2013; 91(1):46–53
3. Yu, Hongyang et al. "A Flexible Hierarchical Bayesian Modeling Technique for Risk Analysis of Major Accidents." Risk analysis : an official publication of the Society for Risk Analysis 37 9 (2017): 1668-1682

Copyright © American Institute of Chemical Engineers. All rights reserved.

550705 A New MILP Model of Multi-Product Pipeline Distribution By Handling Transmix with a Combined Method of Blending, Distillation and Downgrading

Pipeline is much more efficient and reliable to transfer large amounts of petroleum products than other ways via tanker truck, rail cars, and vessel. To save capital and operating costs, one pipeline needs to transfer different types of oil products to different oil depots according to their ordinary demands. As two consecutive different types of oil products inside a pipeline will not be separated by a physical divider, the transmix (TM) will be generated in the interface between oil products. To guarantee the oil quality of product adopted by storage tanks at depots, a considerable amount of TM product from the pipeline needs to be handled appropriately at depots. In reality, TM can be handled by blending, distillation and downgrading. The costs for blending and distillation comes from the storage, blending operation, transportation and reprocessing, while the cost for downgrading is a loss of margin which comes from the different price between the two mixed refined products. Though the downgrade operation and cost is easier and less than blending and distillation, only the TM which consists the same type of products with different grades (i.e., premium gasoline and regular gasoline) could be handled by downgrading the upper grade product to the lower grade one (i.e., downgrade premium gasoline to regular gasoline). Conceivably, the multiproduct pipeline scheduling problem with consideration of handling with TM inside the pipeline present a big challenge to oil industries. There are few works reported in this area (e.g., Rejowski et al. (2003); Cafaro and Cerdá (2004 and 2010)). However, there are still lack of studies on dealing with the TM .

In this work, a continuous-time based MILP model is developed to optimize the transportation and distribution of multi-product, which includes a long-distance pipeline and several depots which consists of storage tanks and transmix tanks dedicated to hold pure product and TM, respectively. The TM generated inside the pipeline will be handled with a combined method of blending, distillation and downgrading. The TM with different types (i.e. gasoline and diesel) will be handled with a hybrid method by blending product-rich mixture into the pure product and distillation of the well-mixed part of the TM; while the TM with the same type but different grades oil products (i.e., premium gasoline and regular gasoline) will be handled by downgrading the upper grade product to the lower grade one. Overall, the scheduling model could provide optimal solutions including injection sequence and volume of product into pipeline, product movements inside pipeline, discharging product to multiple depot tanks, as well as sending out oil product based on consumer demands in the continuous time formulation base. The efficacy of the developed scheduling model is demonstrated by multiple case studies.

Keywords: MILP, Pipeline, Multi-product, Scheduling, Transmix, Blending, Distillation, Downgrading

References:

Cafaro, D., & Cerdá, J. (2004). Optimal scheduling of multiproduct pipeline systems using a non-discrete MILP formulation. Computers and Chemical Engineer- ing, 28, 2053–2068. ISSN 0098-1354.

Cafaro, D., & Cerdá, J. (2010). Operational scheduling of refined products pipeline networks with simultaneous batch injections. Computers and Chemical Engineer- ing, 34, 1687–1704 .

Rejowski, R., & Pinto, J. (2003). Scheduling of a multiproduct pipeline system. Com- puters and Chemical Engineering, 27, 1229–1268 .

Copyright © American Institute of Chemical Engineers. All rights reserved.

550667 Simultaneous Berth and Yard Allocation Planning of Container Vessels at Tactical Level

A reduction in the productivity of port has been observed over the past several years, due to inefficient usage of the available resources. The terminal operator faces the problem of determining berth positions for each of the vessels such that resources can be used as efficiently as possible. One of these resources is the quay, another resource is the yard. Amongst all the operations, the most important is an optimal berth allocation of the ship and yard allocation for the containers. They are inter-dependent on each other and conventional berth allocation and planning models have difficulties to handle the uncertainty of vessel arrival. And terminal operators strive to optimize the berth and yard allocation planning so as to minimize the travel distance of all containers. Because long travel distance not only causes high cost of carriers but also has a potential threat to the loading and discharging operations at the quayside. So, it is crucial to consider these two planning problems in an integrated manner.

To deal with disturbances in transportation schedules, two approaches are getting more and more attention: (i) proactive robustness, which incorporates buffer times into strategic or tactical timetables to deal with disturbances and thus to prevent delay propagation through a well set schedule, and (ii) reactive disruption management, which is concerned with operational recovery after a disruption. The research in this paper focuses on incorporating proactive robustness into the nominal berth and yard planning.

Mathematical optimization has been used to improve the operational capability of port terminals using various realistic constraints. Many papers consider berth planning problems at an operational level. However, only a few papers consider tactical berth planning problems. Furthermore, the simultaneous berth and yard allocation planning problem has rarely been raised, and the associated methodologies have neither been systematically studied. Normally the berth allocation serves as an input for the yard planning problem, and both problems are solved separately. Existing publications are either focused on planning problem without considering realistic flexible arrival and departure time windows or using fixed maximal deviation from nominal arrival time [1].

The goal of this study is to optimally use the resources on the quay. For instance, making the most of quay cranes and minimizing total carrier transportation cost. In this paper, a new time window-based berth and yard allocation planning problem at the tactical level has been developed. Flexible arrival and departure time windows have to be taken into account specifically as control variables. Two different solving strategies: (i) two-stage model, and (ii) simultaneous model are employed to obtain the optimal berthing schedule of each vessel and the optimal schedule of containers transshipment. And the optimization results of these two different solving strategies are presented and compared. Computational results of the case studies demonstrate the efficacy of the developed planning models. Commercial solvers such as CPLEX and ANTIGONE have been employed to obtain the optimal solutions of the developed MIP and MINLP models, respectively.

References

1. Hendriks, M.P.M., E. Lefeber, and J.T. Udding, *Simultaneous berth allocation and yard planning at tactical level.* OR Spectrum, 2013. **35**(2): p. 441-456.

Copyright © American Institute of Chemical Engineers. All rights reserved.

549396 Sensitivity Analysis of Scrub Column Design in LNG Liquefaction Process

In the LNG liquefaction process, there are various methods for achieving the typical LNG specification and preventing the plugging or fouling before entering the cryogenic equipment. Scrub column is one of typical configuration for the heavy component removal. The scrub column removes BTX (Benzene, Toluene and Xylene) and the other heavy hydrocarbons (C5+) from natural gas to avoid freeze-up in the cold end of Main Cryogenic Heat Exchanger (MCHE) and achieve the LNG product specifications such as C5+ Content, Higher Heating Value (HHV) requirement. At meant time, it also provides the feed to Fractionation to generate refrigerant make-up and C5+ condensate product.

The scrub column is often operating near the supercritical region for improving the liquefaction efficiency. Column has huge vapour and liquid traffic difference through the column internals. The BTX and C5+ specification in overhead of column has very strict ppm level purity requirement. So, the column operation is sensitive to the feed gas composition and pressure changes. The current design is heavily dependent on the licensor or engineers' experiences to address those uncertainties and challenges.

There are many literatures related to the sensitivity analysis of different liquefaction technologies. In the presentation, the feed gas composition and inlet pressure changes, ambient temperature variation affecting the column design will be focused. This presentation will describe how sensitivity analysis study can lower the risk of design and potentially reduce capital costs and operational costs. Systematic evaluation of those challenges will provide a path to address and resolve concerns in early stage project. Due to BTX and C5+ ppm level requirement, the accuracy of binary parameters (BIPs) among key components in thermodynamic model also plays critical role for the column simulation and design. The uncertainty of BIPs for column design will also be discussed. Due to the integration of liquefaction plant, the impacts of whole plant performance by those parameters will be presented.

Copyright © American Institute of Chemical Engineers. All rights reserved.

550551 Correlation/Prediction of Swelling of Poly (N-isopropyl acrylamide) Hydrogels in Aqueous Solutions of NaCl Using Extended Enrtl Model

We are establishing a comprehensive thermodynamic model to predict swelling behavior of hydrogels in brine solutions at 298 K. Hydrogels are three-dimensional hydrophilic crosslinked viscoelastic network of polymer chains. Depending on surrounding stimuli such as temperature, pressure, solvent composition, salt content, pH; hydrogels exhibit a significant volume change without changing their structural properties [1]. This interesting property is called swelling. Based on a thermodynamic equilibrium condition developed to describe swelling of nonionic hydrogels in brine solutions [2], Electrolyte Non-Radom Two Liquid (eNRTL) liquid activity coefficient model [3] has been utilized to express activity coefficient of solvents and solutes, and a semi empirical expression is derived [4] from the theory of rubber elasticity [5-6] to illustrate Helmholtz energy of crosslinked polymer network. In addition to solvent-ion interactions [3], interactions of solvents/ions with polymer segments [7], and change in long range columbic interactions due to the presence of polymer molecules are considered in the eNRTL model. To account for configurational change in entropy due to addition of polymer molecules, Flory Huggins lattice theory has been adopted. The correlation model for hydrogel swelling requires only two energy interaction parameters per binary and one network functionality parameter per hydrogel system. No concentration dependent binary or ternary interaction parameter is required. The thermodynamic model has been correlated by regressing experimental equilibrium properties [4] (swelling ratio and solute concentration in the gel phase) of the chemically crosslinked poly (N-isopropyl acrylamide) hydrogels in aqueous solutions of NaCl. Swelling of hydrogels in the solution is controlled by the interactions of species present in the gel. Model correlation results show excellent agreement with data up to NaCl saturation at 298 K and atmospheric pressure. Regressed parameters have been used to predict swelling behavior of similar systems [4] only by adjusting the hydrogel network parameter.

References:

[1] O. Okay, "General Properties of Hydrogels," in *Springer Series on Chemical Sensors and Biosensors*, vol. 6, Springer Berlin Heidelberg, 2009, pp. 1-14.

[2] G. Maurer and J. M. Prausnitz, "Thermodynamics of Phase Equilibrium for Systems Containing Gels," *Fluid Phase Equilib,* vol. 115, pp. 113-133, 1996.

[3] Y. Song and C.-C. Chen, "Symmetric Electrolyte Nonrandom Two-Liquid Activity Coefficient Model," *Ind Eng Chem Res,* vol. 48, no. 16, pp. 7788-7797, 2009.

[4] A. Hüther, X. Xu, and G. Maurer, "Swelling of N-isopropyl acrylamide hydrogels in aqueous solutions of sodium chloride," *Fluid phase equilib,* vol. 240, pp. 186-196, 2006.

[5] P. J. Flory, "Statistical mechanics of swelling of network structures," *J Chem Phys,* vol. 18, pp. 108-111, 1950.

[6] H. M. James and E. Guth, "Simple presentation of network theory of rubber, with a discussion of other theories," *J Polymer Science,* vol. 4, pp. 153-182, 1949.

[7] C.-C. Chen, "A Segment-based Local Composition Model for the Gibbs Energy of Polymer Solutions," *Fluid Phase Equilib,* vol. 83, pp. 301-312, 1993.

Copyright © American Institute of Chemical Engineers. All rights reserved.

550666 A Novel Variable Selection Method for Spectrum-Based Soft Sensor Development

A Novel Variable Selection Method for Spectrum-based Soft Sensor Development

Jangwon Lee*, Jin Wang*, Q. Peter He*+

*Department of Chemical Engineering, Auburn University, Auburn, AL 36849 USA

(JL: jzl0164@auburn.edu; JW: wang@auburn.edu; +QPH: qhe@auburn.edu)

In the last few decades, spectroscopic techniques such as near-infrared (NIR) spectroscopy have gained wide applications in oil and gas industry. As a result, various soft sensors have been developed to predict sample properties from its spectroscopic readings. Because the readings at different wavelengths are highly correlated, it has been shown that variable selection could significantly improve a soft sensor's prediction performance and reduce the model complexity. Existing variable selection methods focus on selecting the variables (i.e., wavelengths or wavelength segments) that are strongly correlated with the dependent variable to improve the prediction performance. Although many successful applications have been reported, such variable selection methods do have their limitations, such as wavelengths selected do not have clear relationship with the chemical bounds or functional groups presenting in the sample. As a result, these methods could face robustness issue and their performances can be highly sensitive to the choice of training data, and deteriorated performance when testing on new samples.

In this work, we present a novel variable selection method that integrate the variable stability and variable importance in the projection (VIP) score and transform them into the probability of variable importance. By incorporating the random selection principle from the genetic algorithm (GA), variables are randomly selected based on the variable importance probabilities, which prevents certain wavelengths never get selected if deterministic variable importance criteria are used. Using several case studies including gasoline and biodiesel, we compare the performance of the proposed method to the existing variable selection methods, including competitive adaptive reweighing sampling (CARS), variable importance in the projection (VIP), genetic algorithm (GA), etc. We show that the proposed method has several advantages: (1) significantly better performance in all case studies; (2) identifies wavelength segments that are clearly related to the chemical bounds and functional groups presenting in the sample; (3) better robustness; (4) fewer parameters and much simpler tuning/training than GA.

Copyright © American Institute of Chemical Engineers. All rights reserved.

AUTHOR INDEX